A SPIT
IS A PIECE
OF LAND

A SPIT IS A PIECE OF LAND

Landforms in the U.S.A.

By DORIS COBURN

Illustrated with photographs

Maps and drawings by WILLIAM JABER

JULIAN MESSNER NEW YORK

Published by Julian Messner, a Simon & Schuster
Division of Gulf & Western Corporation, Simon &
Schuster Building, 1230 Avenue of the Americas,
New York, N.Y. 10020.

Manufactured in the United States of America

Design by Marjorie Zaum

Second Printing, 1979

Library of Congress Cataloging in Publication Data

Coburn, Doris K
A spit is a piece of land.

SUMMARY: Discusses various landforms that are shaped by powerful forces inside and outside the earth's crust.
1. Landforms—United States—Juvenile literature.
[1. Landforms. 2. Geography] I. Title.
GB400.6.C6 551.4 77-17605
ISBN 0-671-32844-1

To

Eric, Gordon, Jeffrey and James

Contents

A SPIT
IS A PIECE
OF LAND

LANDFORMS
Large and Small

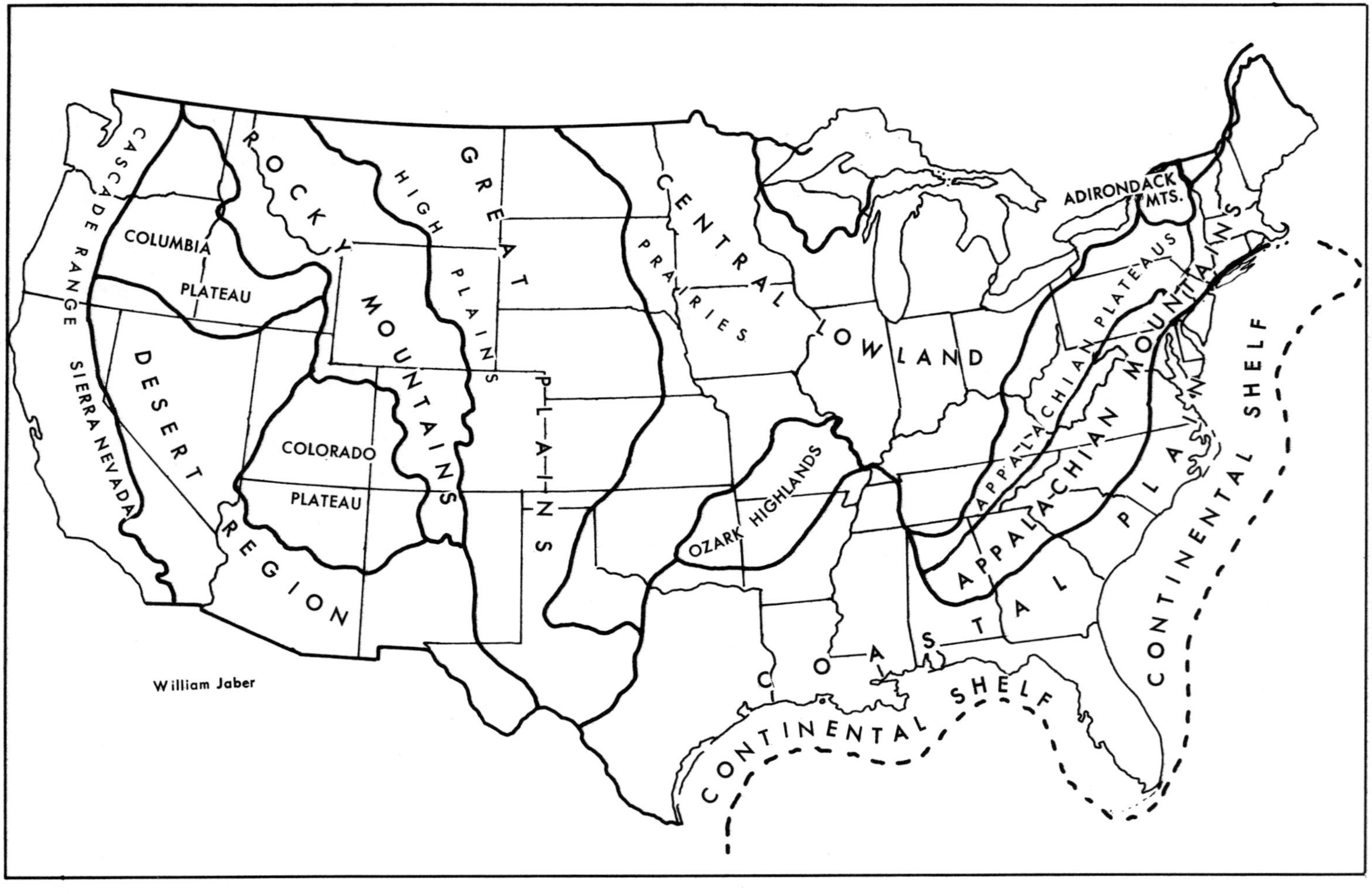
CASCADE RANGE
SIERRA NEVADA
COLUMBIA PLATEAU
DESERT REGION
ROCKY MOUNTAINS
COLORADO PLATEAU
HIGH PLAINS
GREAT PLAINS
CENTRAL LOWLAND
PRAIRIES
OZARK HIGHLANDS
ADIRONDACK MTS.
APPALACHIAN PLATEAUS
APPALACHIAN MOUNTAINS
APPALACHIAN PLATEAUS
COASTAL PLAIN
CONTINENTAL SHELF
CONTINENTAL SHELF
William Jaber

1

Introduction

Landforms

On your way to school, what kind of land do you see? A mountain? A rolling hill? A flat plain? Do you walk between tall buildings so close to each other that they look like sheer canyon walls?

On the surface of the earth, land has many shapes. Like your own face, it has high spots and low ones. Parts are bumpy, others flat. Parts can move up, down, and sideways. From openings, things can be blown out.

The shapes on your face are called cheeks, nose, forehead, jaws and mouth. On the face of the earth, they are called landforms.

The large landforms include *mountains, hills, plateaus*, and *plains*. The smaller ones, and there are many of these, include *spits, deltas, beaches* and *marshes*.

Have you ever wondered if all beaches, mountains and plains are like the ones you know? They are not. Within each landform, there are many variations.

Though landforms differ, all are made on or in the crust of the earth.

Layers Of The Earth

Imagine you are cutting through an orange. You slice through several different layers. The first layer, the skin of the orange, is thin, firm and a little bumpy. A thicker, white, spongy layer is next. Under that is the largest part of the orange, a squashy pulp filled with juice. In the center are hard seeds.

The earth is made up of several layers, too. But the layers of the earth are made of rocks and metals. They differ one from the other in thickness and materials.

The top layer of the earth, the one on which we live, is the earth's crust. It is made of rock, and is about 3 miles deep under the oceans and 40 miles thick under the continents. Yet compared to the rest of the layers, the crust is thin. When you walk or ride, the earth seems quiet. You feel no movement. Yet, the rocky crust is restless. Movement in it never stops. Masses of rock rise and fall, bend, break, and slip sideways.

Under the crust is the mantle, 1800 miles deep. It is a band of hot, doughy, rocklike material full of iron. Pressed by the weight of the crust, the mantle can be squashed or made to spread, somewhat like a piece of putty or soft clay punched by a fist. Some

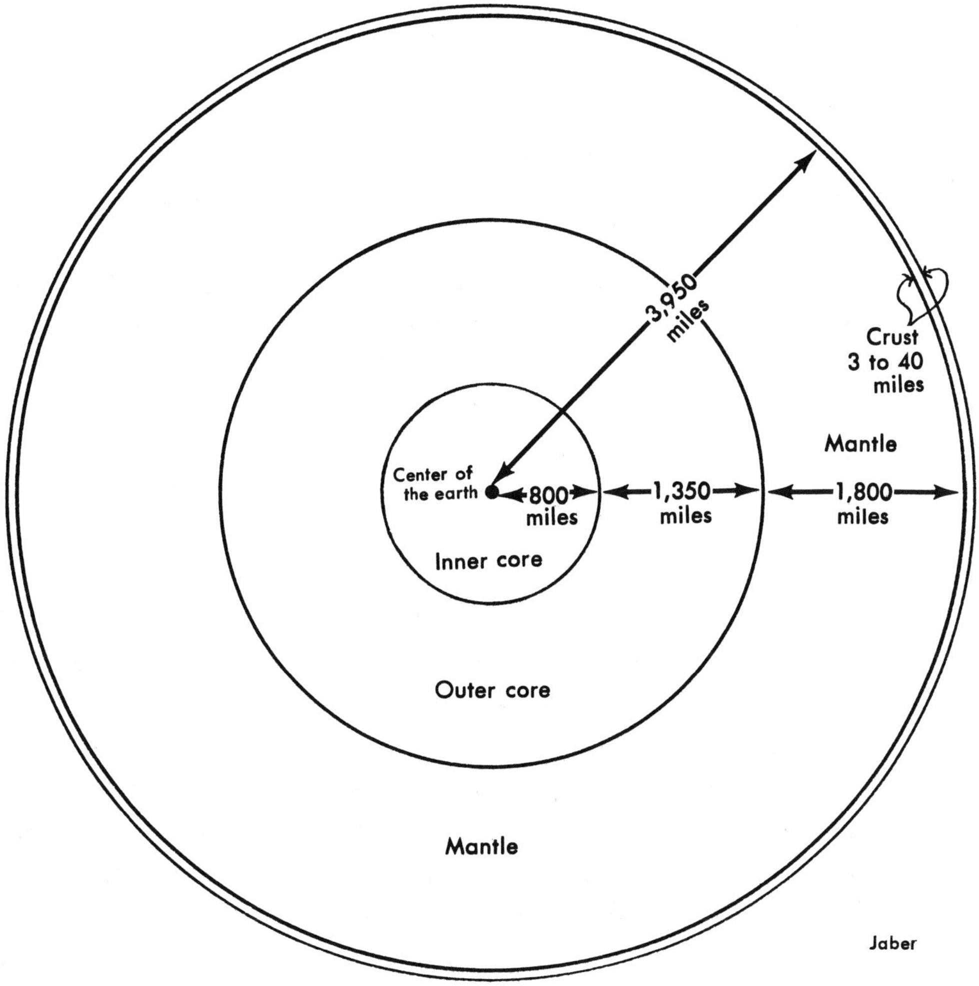

Earth as a ball.

scientists think the mantle may also be made to flow.

The outer core, the next layer, is 1350 miles thick, and is even hotter than the mantle. It consists of iron and nickel in a hot, thick, sticky, liquid state.

In the center of the earth is the inner core. It consists of iron and nickel, too. But in the inner core, these metals form a solid ball 800 miles thick.

Each layer is squeezed by the weight of the layer above it. And the materials inside the layers shift and move slightly because of heavy downward pressure and the rotation of the earth.

Only 500 years ago, most people thought the earth was flat. We know it is a giant ball spinning through space.

Powerful forces from inside and outside the earth's crust shape the land.

Here Today, Gone In A Million Years

Have you ever wondered if the landforms you see every day were always there? If they will last forever?

Landforms don't just happen. They have a life of their own.

All creatures are born, grow, get old and die. Landforms, too, go through stages of growth and change. They, too, are created, have a youthful phase, grow older, get very old, and in millions of years disappear, often to take form some place else.

Two powerful forces bring about these changes. They shape the face of the earth. One force builds up the land, and the other tears it down.

One force comes from inside the earth. It pushes land up from the crust. Through this force, mountains, hills, plateaus, *volcanoes,* and parts of *coastal plains* are created.

The other is a set of forces which work together to break down rock, carry it away and eventually drop it. Usually these forces are called weathering, mass wasting and erosion.

Many of you have seen what happens to a bicycle, a skate or an old car left outside in all kinds of weather. They get rusty and eventually break. Parts may even crumble away.

Have you ever noticed what happens to the paint on a house exposed to weather? After a few years, the paint fades and becomes flaky. Here and there, the paint disappears leaving the wood bare. Some of the exposed wood rots away.

Heat, cold, rain, air, wind and changes in weather are the culprits. These agents of weathering crack the paint and break down the metal. The same thing happens to rock. Weathering agents break up, split, crumble, flake and wear away rock.

In breaking rock, weather and exposure to the air get help from plants and ice. Seeds blown into cracks start to grow. The growing plants force their way up and push the rock apart. Weakened, the rock shatters. If you live in a city or town, you may have seen this happen in pavement cracks. Water, freezing in cracks, expands and splits rock apart, too.

After rock is broken down by weathering, it is carried elsewhere. Loosened rock drops from higher to lower land by gravity. Sometimes it is moved downslope by rain, melting snow or wind. After the fall, the rock breaks into smaller pieces. Moving rock materials downslope is known as mass wasting.

Torrents of water erode the earth.

Erosion, another force in this process, includes weathering, and the transportation and the eventual deposition of the products of weathering. Wind, and moving water in the form of *glaciers*, rivers, waves and currents are the agents of erosion. They wear down the land, and as they move along, they carry rock and other materials from one place and drop, or deposit, them in another place. The weathered material that is transported and deposited is sediment.

Many landforms are described as having been formed by erosion or deposition. *Valleys, canyons, mesas* and *buttes* are formed by erosion. Beaches, *sand bars*, spits, deltas and plains are built up by deposition.

Even though the attack on landforms never stops, the face of the earth changes slowly. Time is figured in millions of years.

Weathering and erosion carved out mesas, buttes and pinnacles in Bryce Canyon.

2

Along the Coast

Coastline

On the map of the United States, you can trace with your finger the line which separates the land from the sea. It is called the *coastline* or *shoreline*.

Only on a map is the coast as thin as a pencil line. Along a continent, the coastline includes all the land reached by the highest storm wave.

Coastlines vary. A coastline may be either:

- straight or irregular, with *capes* and *peninsulas* jutting into the water, coves and bays pushing into the land.
- lined with *cliffs* whose steep, rocky sides plunge into the water or edged by low lying marshes, *swamps*, deltas and beaches.
- bordered by a wide coastal plain or marked off by volcanic mountains sloping into the sea.
- unpopulated or swarming with people and industry.

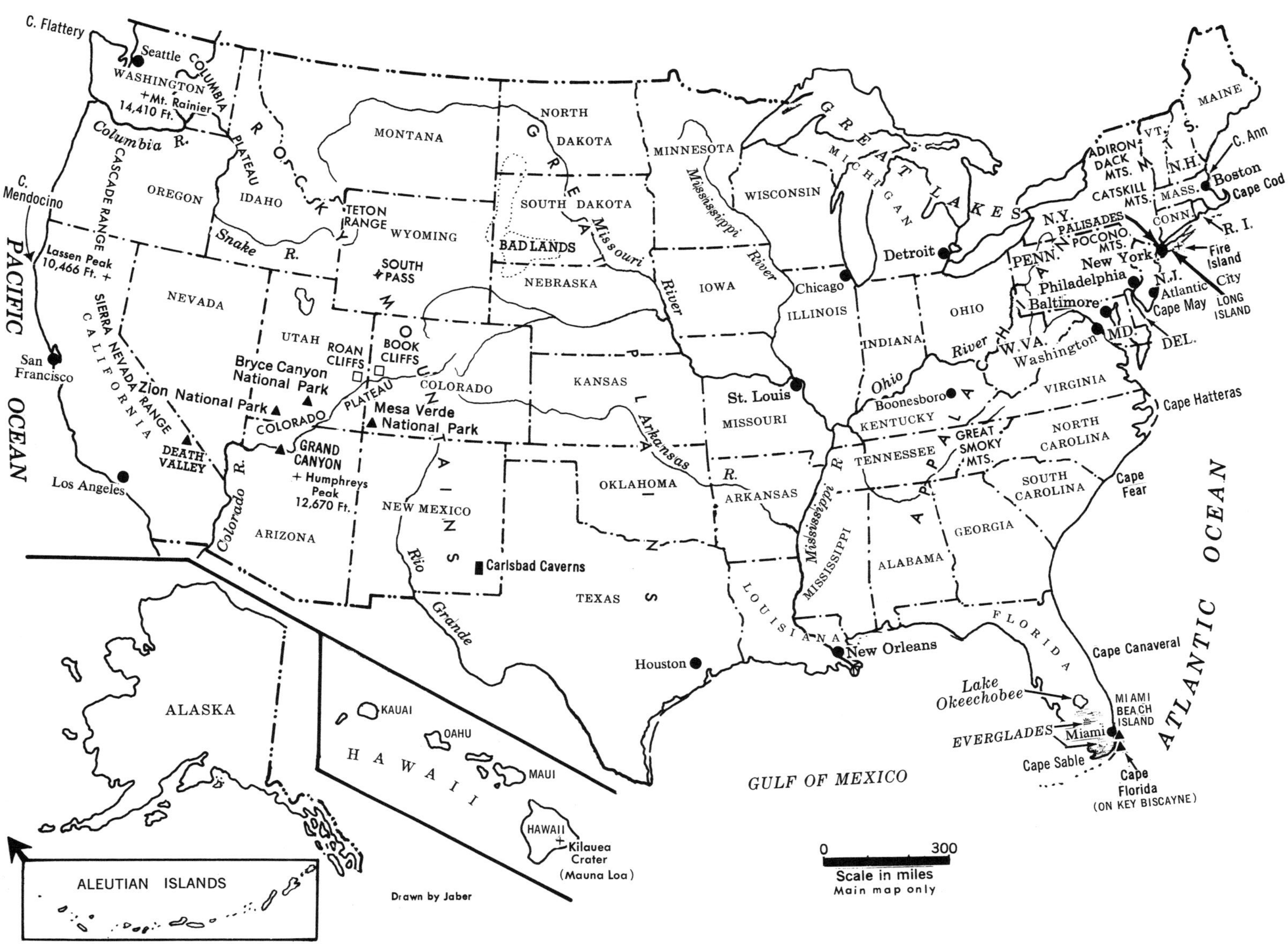

PACIFIC OCEAN
ATLANTIC OCEAN
GULF OF MEXICO
C. Flattery
Seattle
WASHINGTON
+Mt. Rainier
14,410 Ft.
COLUMBIA
Columbia R.
C. Mendocino
CASCADE RANGE
OREGON
Lassen Peak
10,466 Ft. +
San Francisco
SIERRA NEVADA RANGE
CALIFORNIA
Los Angeles
DEATH VALLEY
NEVADA
Zion National Park
Bryce Canyon National Park
UTAH
ROAN CLIFFS
BOOK CLIFFS
COLORADO PLATEAU
GRAND CANYON
+ Humphreys Peak
12,670 Ft.
Colorado R.
ARIZONA
IDAHO
PLATEAU
Snake R.
ROCKY MOUNTAINS
TETON RANGE
WYOMING
SOUTH PASS
MONTANA
COLORADO
Mesa Verde National Park
NEW MEXICO
Rio Grande
Carlsbad Caverns
TEXAS
Houston
NORTH DAKOTA
SOUTH DAKOTA
BAD LANDS
NEBRASKA
GREAT PLAINS
Missouri River
KANSAS
Arkansas R.
OKLAHOMA
MINNESOTA
Mississippi River
WISCONSIN
IOWA
ILLINOIS
Chicago
St. Louis
MISSOURI
ARKANSAS
LOUISIANA
New Orleans
MISSISSIPPI
Mississippi R.
GREAT LAKES
MICHIGAN
Detroit
INDIANA
OHIO
Ohio River
Boonesboro
KENTUCKY
TENNESSEE
APPALACHIAN
GREAT SMOKY MTS.
ALABAMA
GEORGIA
FLORIDA
Lake Okeechobee
EVERGLADES
Miami
MIAMI BEACH ISLAND
Cape Sable
Cape Florida
(ON KEY BISCAYNE)
Cape Canaveral
SOUTH CAROLINA
NORTH CAROLINA
Cape Fear
Cape Hatteras
VIRGINIA
W.VA.
Washington
MD.
DEL.
Baltimore
Philadelphia
PENN.
N.J.
Atlantic City
Cape May
New York
LONG ISLAND
Fire Island
N.Y.
PALISADES
POCONO MTS.
CATSKILL MTS.
ADIRONDACK MTS.
VT.
N.H.
MASS.
CONN.
R. I.
Boston
Cape Cod
C. Ann
MAINE
ALASKA
ALEUTIAN ISLANDS
HAWAII
KAUAI
OAHU
MAUI
HAWAII
+ Kilauea Crater
(Mauna Loa)
0
300
Scale in miles
Main map only
Drawn by Jaber

A straight shoreline. Winds have blown patterns into the sand.

Shorelines are always changing. They are shaped by waves, currents, tides and rivers. There is an ongoing battle between the land and the sea.

Waves, crashing against nearby hills and cliffs, break off large boulders which fall to the shore. The forces of erosion wear away the cliffs and hills until they become smaller and retreat inland.

Large chunks of rock are separated from the mainland and stand alone in the water. In time, these rocky forms are broken down further into stacks, arches and pillars. Eventually these weaken and fall into the sea.

An irregular shoreline, with a headland in the distance. The forces of erosion are breaking down the cliff and pushing it inland.

Stacks and arches are separated from the mainland.

After the land has been flattened, the sea crawls in slowly and covers it. When the sea is the winner, the coastline becomes narrower.

At other times, the land wins the battle. Again, waves, tides, wind and rivers join forces, but this time, it is to build up the land.

Sand is scooped up from the floor of the sea and deposited into shallow waters near the shore. Rivers that start in distant, high places, deposit along the shore the earth, rocks and minerals they are carrying. After shallow waters are filled with sediment, flat, low stretches of land appear. Among them are spits, sand bars, beaches, deltas, marshes and swamps.

Sometimes land is pushed up above sea level from the ocean floor. At other times, the water of the ocean draws back and leaves dry land exposed.

When the land is the winner, the coastline becomes wider.

Coastlines are changed by people, too. Marshes and swamps are drained, filled and turned into farmland, roads and homesites. Beaches are widened. Big dredges scoop sand from the sea and dump the sand onto the shore. Dikes are built to keep water from flooding the land. On beach dunes, grasses are planted to keep the sand from moving away.

The Continental Shelf and the Continental Slope

A continent is one of the big land masses of the earth. We live on the continent of North America. Water surrounds the mainland: the Atlantic Ocean is on the east, the Gulf of Mexico on the south, the Pacific Ocean on the west and the Arctic Ocean on the north.

A continent does not end at the edge of the ocean. The land does not drop suddenly to the bottom of the ocean. But, rather, the land slopes gently away from the coasts to form a platform around the continent. This natural continuation of the coast is the *continental shelf*. Above it, the water is shallow.

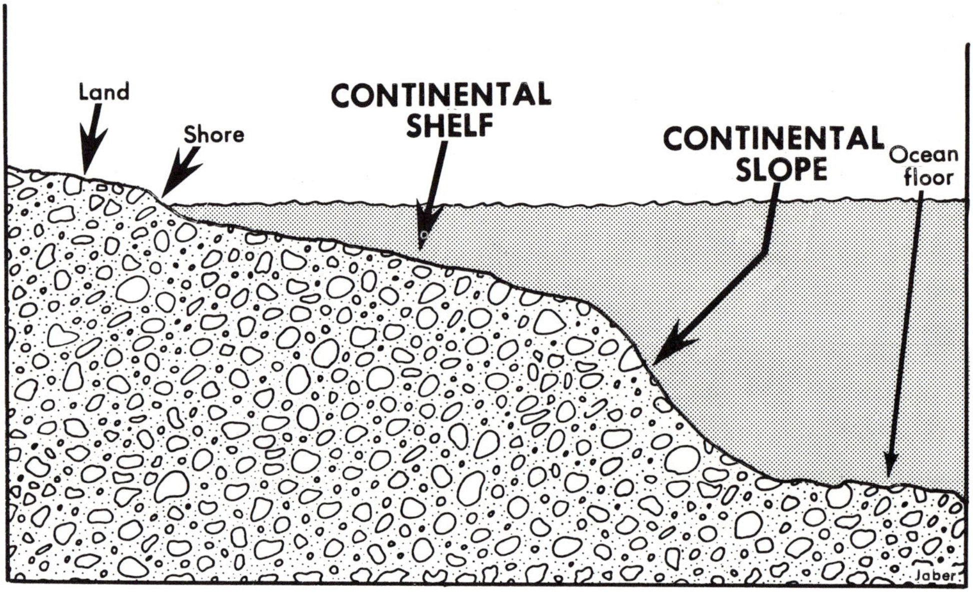

The Continental Shelf and Continental Slope.

Along some coasts, the continental shelf may be up to 500 miles wide. Along others, it is very narrow. In some places, it is less than two miles wide.

From its edge, the continental shelf drops suddenly, like a steep hill or cliff, deep into the ocean. This rapid fall is the *continental slope*. The slope may plunge a few thousand feet, or it may go miles deeper. The continental slope ends at the ocean floor.

Beach

As soon as the days get warm, people flock to the beach to swim, sun, go boating and fishing.

The beach is the edge of land along an ocean, lake or river. The incoming tide washes over the beach and covers it with water. When the tide goes out, the beach is left dry and generally clean, except for a few shells, a bit of seaweed, and maybe some trash, if the water is very dirty.

Beaches differ. They may be either:

- talcum powder soft, sandy, pebbly, rocky or they may be covered with broken bits of lava, shell, coral.
- long or short, wide or narrow.
- bordered by a cliff, a volcanic mountain or by a gently sloping plain.
- pounded by waves or caressed by them.
- washed by icy waters or lapped by warm seas.
- crowded with people or empty with no one around.

But all beaches are formed by materials found near their shores: rock, lava, shells, coral, sand. And all beaches are shaped and reshaped by waves, currents, tides, wind and rivers.

Beaches are part of shorelines. But not all shorelines have beaches.

A long, sandy beach crowded with people, in Massachusetts.

A rocky beach along the shore in Maine.

The beach in Biscayne Bay, Florida, bordered by the coastal plain, is lapped by warm water.

Spit and Sand Bar

A spit may be many things: the saliva from the mouth, a slender pointed rod that holds meat for a barbecue, the image of another person.

As a landform, a spit is also a long, narrow, sandy tongue of land that sticks out into the water. It is attached to the mainland and goes more or less in the same direction as the shore.

Spits form along quiet, sandy coasts. They are built up by ocean currents, waves, sand and wind. Shore currents scoop up sand and other materials from the sea bottom. When waves break along the shore, they lift the sediment and drop it into shallow water. Sometimes the waves drag the sand from the beach and deposit it in the shallow water, too. Over a long period of time, enough sediment piles up to form a spit.

A spit may either:

- extend straight out into the water or the tip may curve toward the shore. A curved spit is called a hook.
- end in open water, grow toward a nearby shore until the two shores meet or connect an island with the main shore.
- be seen at low tide, may be hidden under water at high tide or may be built permanently above sea level.

When a long, narrow piece of sandy land goes in the same direction as the coast and is *not* attached to the land, it is called a sand bar or an *off shore bar*.

Off shore bars are long, narrow, sandy islands. Covered by water, they can be a threat to passing ships. When raised high

A spit with a hook at Long Point and Provincetown Harbor in Cape Cod, Massachusetts.

above sea level, off shore bars develop good beaches and become popular beach resorts. They are known as barrier beaches.

Along the eastern coast, there are a number of well known barrier beaches. Among them are Fire Island in New York, Absecon Island, site of Atlantic City in New Jersey and Miami Beach in Florida. Separated from the mainland by water, these island resorts are reached by boats, bridges, or roads called causeways.

Atlantic City, New Jersey is built on a sand bar.

Dunes

A *dune* is a mound of sand piled up by the wind. The wind picks up the sand, carries and then drops it. The dune forms where the blown sand piles up against a rock, shrub, or even another dune. It keeps growing as more sand is deposited. The strength and direction of the wind and the amount of sand determine the size and shape of the dune. In their travels, dunes may cover a town, a forest or fill a canyon.

To hold the drifting dune in place, grasses and shrubs are planted. The intertwined roots trap the sand and anchor the dune to the spot.

Grasses are planted and fences built to anchor the dunes on Fire Island, New York.

Dunes may either:

- form along beaches, on sandy deserts or on sandy plains.
- be small mounds as high as a car or they may rise as high as a 20 story house.
- move short distances or travel for miles.
- cover large areas or small ones.
- have steep slopes or gently curving ones.
- take the shape of moon crescents, curved sword blades or stars.
- have smooth tops or ones rippled like waves.
- be made of different kinds of sand, among them quartz and gypsum.

But all dunes develop in flat places where sand and wind are plentiful, and where grasses or other kinds of plants are scarce.

Peninsula

Your nose is like a peninsula. It is attached in one place and pokes out.

A peninsula is part of the mainland that juts into the water. Surrounded by water on three sides, most of it is coastline. Peninsula means "almost an island."

Peninsulas may be wide or narrow, long or short. There may be many peninsulas in a state, the state itself may be a peninsula, or there may be peninsulas poking out of a peninsula state.

The state of Florida is a peninsula, 400 miles long. The waters

Dunes march over a forest.

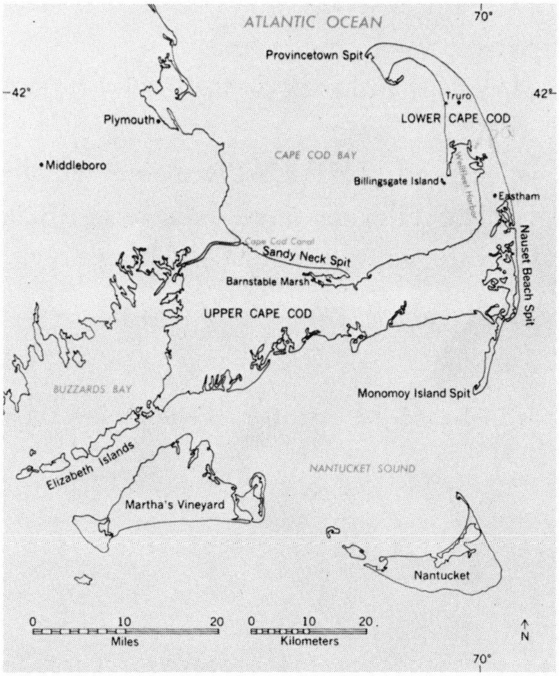

Cape Cod, a sandy peninsula, juts into the Atlantic Ocean like a crooked arm. Spits, marshes and islands are part of the landscape.

surrounding it on three sides are the Atlantic Ocean and the Gulf of Mexico.

Alaska is a peninsula, too. It is attached to Canada on one side. On its other three sides are the Arctic Ocean, the Bering Sea and the Pacific Ocean.

Cape

A cape is a point of land that juts into an ocean, lake or river. The word cape means "head" or "cap." It has many names: point, headland, promontory. All peninsulas end in a cape.

A cape may be either:

- a rocky cliff, or a flat piece of land.
- surrounded by rocks and hidden sand bars dangerous to ships or bordered by deep waters that make good, safe harbors.
- a city full of people and industry, a quiet village, a vacation resort or a sparsely inhabited place.
- be built up by sand, gravel and rock deposited by rivers, waves and currents or by glaciers.

To people living in New England, THE CAPE means usually Cape Cod in Massachusetts. The Pilgrims landed there in 1620. Today, Cape Cod is known for its summer resorts, fishing towns, dunes and artists' colonies.

Many miles of Pacific Ocean shoreline can be seen from Cape Foulweather, → Oregon, a rocky headland.

To the astronauts, THE CAPE means Cape Canaveral, on the east coast of Florida. From that point of land, astronauts were rocketed into space.

A glance at a map of the United States will show many other capes poking into the Atlantic Ocean. Among them are Cape May in New Jersey, Capes Hatteras, Lookout and Fear in North Carolina and Cape Sable at the tip of Florida. Along the Pacific Ocean, there are Cape Slattery in Washington and Cape Mendocino in California.

Cape Lookout, North Carolina, is a flat piece of land.

3

Lowlands

Inland Plain

The floor of your classroom is level. The schoolyard is almost if not quite level. Can you imagine land as flat as the schoolyard but stretching as far as the eye can see? Such a broad, nearly level stretch of open land is a plain. Plains are always lower than the land around them. They are formed by layers of rock material deposited by rivers, glaciers and volcanoes.

In the United States, there are two large *inland plains*. One, the Central Lowlands, is a low, rolling plain covered in most places by glacial deposits. The other are the Great Plains. Together they stretch from the Rocky Mountains to the Appalachian Mountains.

Millions of years ago, the Great Plains region was the floor of a shallow inland sea that had emerged. The floor was covered with sediment dropped by rivers that flowed from the Rocky Mountains. In time, the plains were built up.

Now, the Great Plains are broad, almost flat, grasslands. Often they are treeless. Without trees to stop them, the winds blow fiercely over the land. The Great Plains are divided into two parts: the western or High Plains, and the eastern section called the prairies.

Although the Great Plains appear level, the western part slopes upward toward the Rocky Mountains, reaching a height of over 4,000 feet above sea level. Most of the High Plains are dry. In some parts, sudden rainstorms, sand and wind have carved eerie shapes out of the rock. One such place is called The Badlands of South Dakota, a fitting name for this desolate area. Violent rainstorms have dug deep gullies in The Badlands. In time, the

The Badlands of South Dakota.

forces of erosion will make this section as flat as the rest of the Great Plains.

On the prairies part of the Great Plains, stretches of treeless land are covered with deep, fertile soil which gets plenty of rain. The corn, wheat and rye grown there are used for making bread all over the world. The land is called the "Bread Basket of the World."

In the early days of American history, when the Indians lived on the plains, they moved from place to place. And so the grasses on which their animals grazed were not used up.

However, in modern days, ranchers permitted too many animals to graze in one area for too long a time. Every season, farmers planted wheat, corn, cotton and other crops in the same fields. The land was not given a chance to rest and get back its strength. The soil lost its richness, and only a thin layer of good topsoil covered the sand beneath.

One year there was little rain. Great winds blew away the top soil, leaving only the sand. Then the sand was blown over the land until it was heaped on roads, against houses and barns. Much of the plains became a dust bowl. The land was eroded by drought, wind and poor farming methods.

Today, farmers protect the soil so that it will not be blown away again. They rotate crops, terrace the land to keep the water from running off, use fertilizers and give the land a resting period.

Drifts of sand surround a Kansas farm.

Coastal Plain

In the United States, there are two coastal plains: the Atlantic and Gulf coastal plains.

The Atlantic coastal plain extends from New Jersey to Florida. Part of the plain is the continental shelf which was lifted from the ocean millions of years ago. Some of the plain is formed by sediment deposited by rivers, waves and currents.

The Gulf coastal plain, considered part of the Atlantic coastal plain, extends from western Florida to southern Texas. It was formed by sediment deposited by the Mississippi and several other rivers.

The coastal plains are the zone of low, level land between the

seacoast and the nearest hill. In some places, the plains are so flat that water collects on them, causing marshes and swamps to form. In other places, the land rises several hundred feet.

A coastal plain may either:

- extend inland for short distances or spread over several states.
- have regular or irregular shore lines.
- be widened when the continental shelf emerges, when sediment is deposited along the shore or it may become narrower when the ocean drowns the land.

Large cities are located along coastal plains. Many people live there. The soil is fertile, the harbors good and transportation easy.

Orange groves in the low, flat coastal plain of Florida.

Delta

A delta is a plain formed by sediment deposited at the mouth of a river. The delta of the Mississippi River is considered part of the Gulf coastal plain.

A river picks up rock, sand, silt and soil as it moves along. At the end of its journey, it drops the load it has been carrying. When the river flows into a quiet gulf, bay or sea, layers of sediment build up.

After thousands of years, the deposits build up and get so thick, they block the mouth of the river. Seeking its way to the open water, the river moves away from the blocked mouth. As it changes direction, the river divides into one or more branches called distributaries. Each distributary continues to flow away from the main river. All this time, sediment is dropped and spread between the branches.

When each of the newly formed channels is blocked, the distributary shifts and splits again. More and more deposits are spread about. After the blocking, shifting and depositing is repeated many times, a delta is formed. The delta continues to grow toward the sea. As it grows farther into the sea, the delta widens like an open fan.

Like Delta, Δ, the fourth letter in the Greek alphabet, some deltas suggest the shape of a triangle. But not all. The delta of the Mississippi River looks like the foot of a giant bird.

The Mississippi River delta was built up where the river empties into the quiet waters of the Gulf of Mexico. It took about a

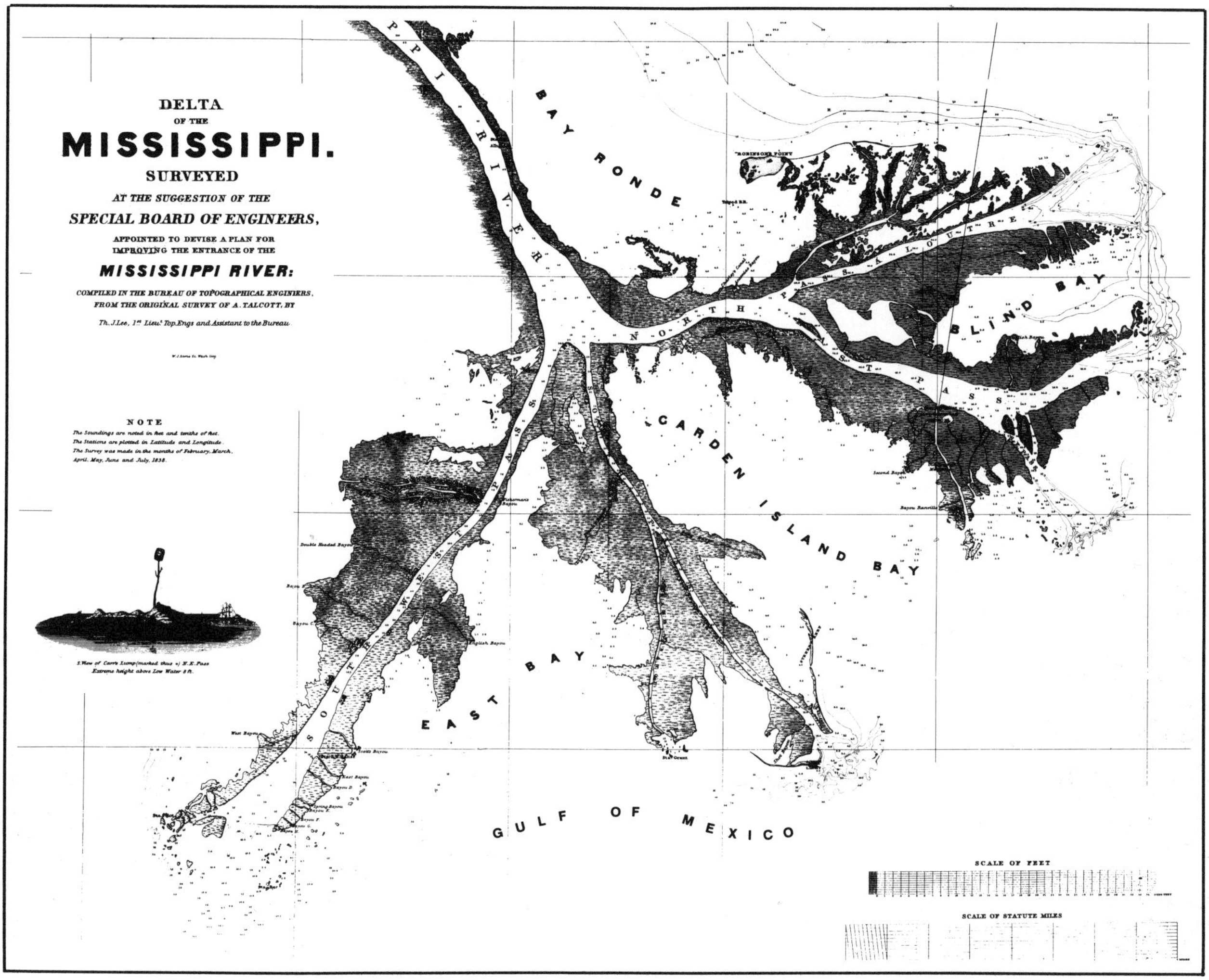

The Mississippi delta takes the shape of a bird's foot. Note the distributaries in each section of the foot.

An aerial view of the southern end of the Mississippi delta where the river flows into the Gulf of Mexico.

million years for the Mississippi to make its delta. Each year, the delta grows about 350 feet. Today, it extends about 200 miles into the gulf, and it is still growing.

The Mississippi River built one of the largest deltas in the world. But, of course, not all rivers build deltas.

Marsh and Swamp

Marshes and swamps form where the land is so flat that water collects on it. The two words, marsh and swamp, are interchanged often, but they differ in many ways.

A marsh is a wetland where tall grasses and water-loving plants grow. Wild birds, insects, animals and wild flowers make their homes there, too. For part of the year, the low-lying land is covered with a shallow sheet of water. The water is either slow moving or it just stays there, unable to flow off.

A swamp is a wetland where trees and shrubs grow. Water collects in the low-lying, wet, spongy ground and cannot drain off or flow away. The land is below water level.

On the surface, swamp water is covered with a green scum and with thick mats of weeds. Here, mosquitoes and insects breed. They provide an early link in the food chain for the larger creatures of the swamp.

Swamp land is too wet for growing fruits, grains or vegetables. But trees and shrubs thrive. Their dense, tangled undergrowth is a refuge for insects, fish and animals. Their tops are resting and nesting places for birds.

A marsh in Georgia.

Okefenokee Swamp in Georgia.

Marshes and swamps may be either:

- found inland near lakes and rivers or along seacoasts, often behind sand bars and spits.
- filled with fresh water of rivers or salt water of the oceans.
- old or newly formed.
- small as a classroom or may cover large areas.
- formed by silt and mud dropped by a river or built along the coast by incoming tides.

Both swamps and marshes undergo change.

When a pond or shallow lake is filled in with tall grasses, a marsh is born. When the grasses cover the water, a meadow may form. When shrubs and trees grow in a marsh, they crowd out the grasses. Leafy tree branches deprive the grasses of the sun, and the grasses die. When the dead plants and silt fill in the marsh, a swamp is born. If, in time, the land were to rise above the water, or if the water were to be drained away, the swamp would vanish. A flat, low piece of land would appear.

Wetlands are changed by people, too. They reclaim the land by draining or filling it with earth, tin cans or old cars. Then they build farms, roads, factories and airports.

After wetlands are drained, birds, animals, fish and plants die. To protect the wildlife and to keep the wilderness in its natural state for all to enjoy, our national and some state governments are buying wetlands for parks.

One of these parks is the Everglades National Park in the southern part of Florida. It was established in 1947.

The Indians who first lived there called the region "*Pay-hay-o-kee*," which means grassy waters. They named it for the fields of stiff, tall, saw-leafed grass that stretch as far as the eye can see.

The Spaniards who explored these lands described the plains of saw grass in their reports. They also wrote about the silence, the loneliness, and the mystery of the marshy wilderness. Impressed with its haunting spell, they called it *El Laguno de Espiritu Santo*, The Lagoon of the Holy Spirit.

Later, when the English explorers arrived, they referred to it on maps simply as the *Glades*.

The Everglades is a unique land form found only in the United States. Overflow water from Lake Okeechobee in the central part of Florida, spreads out across the land in a thin sheet of water. It flows south, very, very slowly until it enters Florida Bay. The water is covered with broad fields of saw grass, and is often referred to as a *river of grass*.

The flow of water passes through marshes, swamps, lakes, bays, and rivers on its way to Florida Bay. However, some dry land in the Everglades rises about seven feet above sea level.

Wild life abounds. Leggy wading birds reach through the water of the marshes to stand on the soft earth beneath. They lift up fish for their meals. Alligators, turtles, frogs, fish and shrimp share the sluggish waters. Colorful birds nest among the trees, and often come to the water to feed. Racoons, turkeys, pumas and snakes live throughout the Everglades.

Dotted here and there on the plains of saw grass are small islands of trees called hammocks. They rise slightly higher than the fields of grass. A ditch or channel of water surrounds each hammock. Clumps of mangrove, cypress and pine trees root in the muddy marshes. Silt and sediment are trapped by their roots. In time, a hammock builds up.

The Everglades National Park is open all year. Visitors follow well marked trails, stroll along boardwalks built above the swamps and marshes, and travel in small boats over watery courses to view the wonders of the Everglades wetlands.

SCENES FROM THE EVERGLADES NATIONAL PARK.

Ospreys making their nest.

← Dense forests of trees.

Alligator resting.

4

Highlands

Cliff

A cliff is any high, steep face of solid rock that goes straight up. Some cliffs rise from the shore of a sea, a lake or a river. Some rim desert floors. Others form the sheer rock walls of canyons.

Along the New Jersey side of the lower Hudson River are the majestic Hudson River Palisades, cliffs that rise high above the river bank. Looking like massive walls of a fortress, the Palisades stretch for about 30 miles along the shore. During the Revolutionary War, the American army built Fort Lee on the flat top of this cliff. From that position, the Americans had a commanding view of the countryside. They could watch the British ships on the river, and follow the movements of troops on the land.

Palisades along the Hudson River.

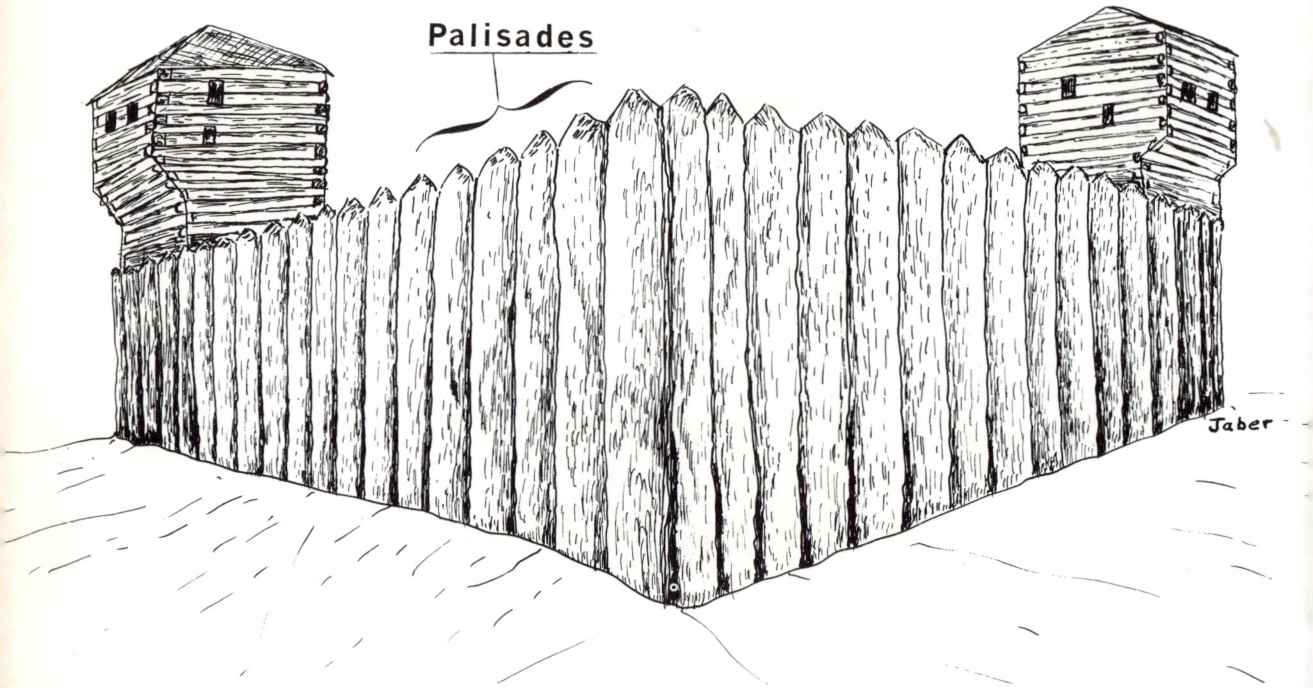

Palisades originally meant a wall or fence made of upright, pointed posts, set close together, usually for defense. Early American forts were often built with palisades as walls. Rock cliffs, such as the Palisades of New Jersey and New York, have many vertical cracks, and so from a distance look like a giant palisade wall.

In the southwestern part of the United States, in regions that are hot and dry, there are cliffs that rise above canyon floors. Long before the Spanish conquerors explored that part of the country, a tribe of cliff dwellers, the Indians of the Four Corners, lived there. The homes they built on the ledges of the cliffs were designed like apartment houses, with many stories and rooms.

Early homes of the cliff dwellers. The canyon cliffs are pitted with caves.

Though the homes remain, the tribe disappeared. No one knows where they went or why.

For many years, visitors to the site took souvenirs to show friends back home. To protect and preserve these ancient cliff dwellings, the government established Mesa Verde National Park in Colorado.

Like other land forms, cliffs are attacked by the forces of nature. At the edge of the ocean, cliffs are pounded by waves until some of the rock breaks off. Along river banks they are broken down by weathering, too. Rock on the surface crumbles. Cracks on the side and the top of the cliff get bigger until some of the material breaks off and falls to the base in the form of sand, pebbles, chunks of rock, and even boulders. In hot, dry country, cliffs are eroded by wind and running water.

Eventually, they are worn back and may, in time, disappear.

Hill

Rivers flow downhill. The steeper the hill, the faster the flow of water.

You start to climb a hill from its base. You go up its sides or

← In Maine, the cliffs rise out of the sea.

slopes. When you reach the top, you are at its summit. From their summits, hills provide a good view of the countryside.

Hills are pushed up from the crust of the earth, and rise naturally above the land surrounding them. Smaller and less rugged than mountains, their slopes are more gentle. After millions of years, some mountains are worn down into hills.

A hill may either:

- rise slightly, gradually or steeply from its base. A very low hill is sometimes called a knoll or mound.
- have a rounded top called a dome or a flat one known as a knob.
- stand alone or may be connected by ridges to other hills.

Mountain

A mountain stands high above the land around it. But what is called a mountain in one part of the country might be called a hill in another. For example, the Black Hills of South Dakota are higher than the Appalachians or any other mountains in the eastern United States.

To be a mountain, the earth's surface must rise abruptly at least

A hillside in Vermont on a still summer's day. →

2,000 feet. Mountain walls are steep, and sometimes they go straight up and down.

Scientists disagree on the beginnings of mountains. However, some of them point out that most mountains found near the edges of the North American continent go in the same direction as the shoreline. These scientists believe mountains begin at the bottom of the sea.

Material is deposited along the seacoasts, and piles up in places where the crust is sinking slowly. The piles grow several miles thick, and a few hundred miles to a thousand miles wide. It takes hundreds of millions of years for this process to take place.

Forces in the mantle, which is under the crust, pull the pile down. One layer of sediment presses upon the one beneath it. Packed down by the weight of the material, the sediment changes into rock.

The heavy load presses on the crust, and causes it to sag. One side pushes against the other. Squeezed, the sagging crust folds up like an accordion. Tremendous pressures inside the earth slowly push up the crust. The rocks are uplifted, and folded into a chain of mountains. The once sagging crust returns to its original level, leaving part of the mountain above the earth's crust and part below.

The Appalachian Mountains in the eastern part of the United States are folded mountains. They have long folds, each one lying parallel to the other.

A view of the Grand Teton, a block mountain. →

But mountains don't all look alike. Nor are they formed in the same way.

The Teton Range, in the state of Wyoming, is an example of block mountains. This kind of mountain is born when great pressures cause the crust to break. Huge, solid blocks of rock are lifted

above the surface. Some blocks go straight up. Others tilt, and have a gentle slope on one side and a steep cliff on the other.

Then there are some highlands that are called "mountains." They look like mountains, but they are really cut up plateaus. The Pocono "Mountains" of Pennsylvania and the Allegheny "Mountains" of West Virginia are examples.

Mountains pass through stages.

Young mountains, like the Rockies and Sierra Nevada, have:

- high, sharp, jagged peaks.
- rocky, rugged, steep slopes.
- snow-covered summits.
- bald spots of shiny rock.
- swift, down-cutting streams.
- narrow valleys and steep canyons.
- grinding glaciers traveling slowly down their slopes.

A plateau is cut up to look like "mountains."

The gentle folded slopes of the Great Smokies, one of the oldest mountains in the United States.

Older mountains, like the Appalachians and the Adirondacks—all in the eastern part of the United States—have stopped growing. Erosion is wearing them away. They have:

- low peaks and rounded surfaces.
- gentle sweeping slopes covered by dense forests.
- wide valleys.
- slow-moving rivers that flow over the valley floor.

Very old mountains have been worn down to almost a plain. After millions of years, they will disappear, leaving only their roots.

Mountains look strong, hard and unchanging. Yet, they are broken down as soon as they appear. While mountains are being worn down in one place, new ranges rise slowly in another.

Volcanoes are mountains, too. They are formed by volcanic action.

Volcano

Have you ever kicked a rock? Picked one up and thrown it? The rock felt solid and hard. Yet that solid rock can be melted and made to flow like toothpaste coming out of a tube.

There are places deep in the earth's crust where the heat is great enough to melt rocks into a thick, syrupy liquid. Below the ground, the molten material is called magma. At the surface, it is lava. After it cools and hardens, it is igneous rock. (Igneous means "of fire.")

When the molten material is pressed, it flows to the surface through weak spots and openings in the crust. It may either flow slowly, quietly to help form plateaus and palisades, or it may be blown out violently with explosive force, and eventually form a volcanic mountain.

The openings or vents in the crust are called volcanoes. The hills and mountains formed by the discharged materials are called volcanoes, too.

When a mountain volcano is born, pieces of solid and molten

rock, volcanic ash, hot liquids, and fire burst forth into the air. Charged with gases, the volcanic materials burn and explode like firecrackers. Lava flows. Some of the dust, ash and cinders fall back and form a mound around the opening in the crust. The flowing lava mixes in, and helps build the mound. After each eruption, the volcanic materials pile up on the mound, and also spread out in all directions. In time, a cone-shaped mountain with steep sides builds up. Each time there is an eruption, the cone gets higher and wider.

Shishaldin Volcano, a cone volcano, rises 9,372 feet above sea level on Unimak Island in the Aleutian chain in Alaska.

At the top of most volcanoes is a depression called a crater. Eruptions usually occur through the crater.

A volcano may either:

- have one crater or many.
- appear in groups as in the Hawaiian Islands, in long chains as off the coast of Alaska, or stand alone (Mt. Hood, Oregon).
- rise out of the ocean (Virgin Islands in the Caribbean Sea), or rise from the land (Mt. Rainier in Washington).
- appear suddenly almost full grown or take years to develop.
- occur where earthquakes are common, erupt after an earthquake or have no connection with earthquakes.
- be active, dormant or extinct.

A volcano is considered active while it is erupting (Mt. Kilauea in Hawaii). If it has stopped erupting and has slept during modern times, it is dormant (Mt. St. Helens, Washington). When all activity has stopped, a volcano is said to be extinct (Humphrey's Peak, the highest point in Arizona). However, dormant and extinct volcanoes have been known to erupt. Mt. Lassen in California was thought to be extinct, but suddenly, in 1914, it became active.

Plateau

A plateau is a high, flat section of land that covers a wide area. It usually stands higher than the surrounding country on at least one side, as in the Appalachian Plateau. Its edges are sometimes steep and cliff like. Its surface is severely eroded into canyons and valleys. A few plateaus are bordered by high, rugged mountains on at least one side, as in the Colorado Plateau. Sometimes, a plateau stretches like a flat table between mountain ranges, as in the Columbia Plateau.

In its youth, a plateau has a high, broad, flat top. A few rivers gush over it, and cut out narrow box like *valleys*. These are known as box canyons. Later, these canyons become deeper and wider.

As it grows older, the plateau is cut by many rivers. The rivers cut valleys that separate the plateau into sections. In dry, hot country, the separate sections of the once broad, flat top of the plateau are reduced to the small, flat, narrow tops of mesas, buttes and pinnacles. In other parts of the country, the cut-up plateau looks like mountains. And the name "mountains" is used, such as the Catskill "Mountains" in New York State.

The Catskill "Mountains" of New York State are a cut-up plateau.

When a plateau is very old, it is worn down to the level of a plain with a few hills standing here and there.

Plateaus may form when masses of rock are uplifted from the earth's crust. The Appalachian Plateau is uplifted. It rises about 2,000 feet above sea level. Forests and cultivated fields cover this plateau.

The Colorado Plateau is uplifted, too, and rises from about 5,000 feet to about 10,000 feet above sea level. The region is hot and dry. Many streams have cut down into the plateau so that in places it is deeply eroded. Canyons, mesas, buttes, spires and other fantastically shaped rocks are common. These landforms may be seen in national parks located in the region. Among them

An aerial view of the Colorado Plateau near Utah. The meandering river was formed before the plateau began to rise.

are Canyonlands, Bryce Canyon, Zion, Mesa Verde, and Grand Canyon National Parks.

The Columbia Plateau reached its heights by volcanic activity. Slowly, quietly, from many openings in the earth's crust, lava poured out. It traveled great distances, filled in valleys and covered peaks. Layer upon layer of lava built up the land to a height of 5,000 feet. Lava cones and volcanic scenery are features of the Columbia Plateau.

Mesa, Butte, Pinnacle

Mesa in Spanish means "table." It was given its name by the Spanish explorers of the southwestern part of the United States. A mesa is a large, flat topped hill with steeply sloping sides. It stands alone, and rises high above the land surrounding it. Many are found in hot, dry regions.

A flat topped mesa is formed from a flat topped plateau. Streams cutting through the old plateau separate the mesa from it. In its old age, the plateau is worn down until it is flat. The mesa, standing high, is a reminder of the once high plateau.

Hard rock covers the top of a mesa like a cap. The hard rock top keeps the soft rock under it from crumbling away as quickly as the rest of the plateau. But in time, pieces of rock break off from the top, fall to the ground and are carried away. The mesa becomes narrower. When the mesa is worn down and takes up less space, it is called a butte.

A butte, too, stands alone and rises abruptly from the land. But

Scotts Bluff Mesa in Nebraska is part of an old plateau.

its sides are steep and craggy. Its top may be either flat, rounded or pointed.

Erosion attacks the buttes, and sculptures them into tall, thin pinnacles. In time, these forms are worn away further. At the end, the little pieces of rock left standing are blown into bits by the wind and washed away by the rain.

5

Valleys

Valley

Try this. Press a finger hard in the middle of your palm, or cup your hand as if you were catching a ball. The long crease between the sides of your hand is a "valley."

A valley is a long stretch of land lying between hills or mountains. The bottom of the valley is its floor; the sides are its slopes or valley walls. A river usually flows along the floor.

River valleys are started by swift-flowing streams that cut into the tops of mountains or plateaus. Valleys are cut by gravel, sand and other sediment carried along in streams. This material scrapes the sides and bottom of a stream until a narrow V-shaped trench forms. This trench is the beginning of a valley.

When a valley is still small, it is known as a gully. When it grows a little larger, it becomes a ravine or gulch. As the river wears away the land, the depression gets deeper and deeper. At the same time, as the river scours the floor, it also breaks chunks of rock from its banks, making the valley trench wider. In the hot, dry western lands, a young valley is called a canyon. In other regions, it is a chasm or gorge. Waterfalls and rapids may rush over their narrow floors.

With the help of other forces of weathering and erosion, the valley walls continue to wear away. Erosion starts on the upper parts of the walls. Layers of rock are stripped back. The slopes become gentler, and the walls move apart, keeping the V-shape the valley had when it was just a trench.

As more and more material is removed, the valley becomes lower, flatter, wider, and starts to get a U-shape. In time, the mountain is separated in two, and its sides become the borders of a valley. What was once a ditch or gulley is now a young U-shaped valley.

As the valley grows older, the river cuts into the land, captures nearby smaller streams and gets longer. With more water, it can

← Oneonta Gorge in Oregon is a young valley. It is bordered by sheer, moss-covered cliffs, and waterfalls and rapids flow along the floor.

carry heavier rocks and more sediment. Now, instead of cutting down, the river begins to cut the valley's walls more quickly than the bottom. It swings from side to side in snakelike fashion, breaking down its banks. This action occurs when the river begins to

The older valley widens: Ojai Valley in California (below) and Shenandoah Valley in Virginia (opposite).

carry a heavier load of sediment. At the same time, it deposits some of this sediment and builds up a plain. When the plain is so wide that the stream's bed does not take up all the land, the valley is "mature."

A mature valley becomes an old one when the river reaches the base of the mountain or plateau, and moves lazily over the valley bottom it carved out. Downcutting stops, but the forces of erosion continue to cut sideways. Sometimes the sides of a valley are so far away from each other, they cannot be seen except in the distance. Such valleys usually are dotted with large cities and small towns and villages. Valley bottom lands have the richest soil in the world.

Along the coast, an old valley may be drowned by the ocean. The valley floor disappears under water to become the floor of the bay.

Canyon

Canyons are often called "upside down mountains." Mountains rise from the surface, but canyons go deep into the earth from the flat rims of plateaus.

A canyon is a deep valley with high, steep sides. It usually forms in deserts and on high plateaus where the air is hot and dry. A swift river flowing over a young plateau starts the formation of a canyon. It cuts through the rocks like a sharp knife. Rapidly it slashes to a great depth. The river then flows at the base of the canyon between the walls it carved out. Rock seems to be everywhere.

A young canyon has towering walls that go straight up from a narrow floor. The top is almost as wide as the bottom. Weathering is slow and the tops of the walls remain close to each other. A swift river, waterfalls and rapids rush over the floor, cutting it deeper and deeper.

As it grows older, the top of the canyon walls break off and fall down. The forces of erosion scoop out large chunks from the sides.

A young canyon.

They strip away layer upon layer of rock. Rivers and rain wash away loose rock. The canyon walls move away from each other, and the canyon opens up wide. The river flows along the floor and continues to cut deeper and deeper into the earth.

In old age, the walls of the canyon are very far away from each other. Mesas, buttes and pinnacles, separated from the plateau, stand alone on the canyon floor. After millions of years, the canyon is worn down to a plain.

In the United States, the greatest of all canyons is the Grand Canyon of the Colorado River. It was carved out across the Colorado Plateau in the states of Arizona, Utah and Colorado by the Colorado River. This majestic canyon is 217 miles long, 7 to 15 miles across from rim to rim and a mile deep. It took millions of years to create, and it continues to grow wider and deeper.

The Grand Canyon is remarkable not only for its size, but also for the varied colors of the many exposed layers of rock. Each layer tells a part of the earth's story.

Theodore Roosevelt, 25th president of the United States, said that the Grand Canyon of the Colorado is "the one great sight every American should see." Today, it is part of the Grand Canyon National Park, established in 1919, and the Grand Canyon National Nature Monument, established in 1932.

The Colorado River cuts through the horizontal layers of the Colorado Plateau → to form the Grand Canyon.

Pass, Water Gap, Wind Gap

In 1803, the United States bought from France the entire Mississippi Valley, and a lot of the land west of it, right up to the Rocky Mountains and beyond. The transaction, which doubled American territory, is known as the Louisiana Purchase. The following year, President Jefferson sent the explorers, Meriwether Lewis and William Clark, westward to find out how Americans could best cross this land, settle and develop it.

When the explorers returned, they reported there was no way to get wagons over the high Rocky Mountains. So, many people gave up any idea of moving west. But in 1824, a small band of fur trappers accidentally discovered a low notch or mountain pass that led right over some low ridges. Its trail was wide enough to get wagons across. They named it South Pass.

Within a few years, thousands of settlers and their wagons were moving across the Rocky Mountains to Oregon and California by way of South Pass. Without that pass, located in what is now Wyoming, the opening of the West would have been delayed many years.

A *pass* is a break or gap in a mountain range through which a road or trail can be built. It forms a passageway to the other side of the range. Many passes in eastern United States are known as gaps. A gap is a break or notch in a mountain that has been cut by a river. If the river still runs through the gap, it is a *water gap*. A well known one is the Delaware Water Gap, at Stroudsburg, Pennsylvania. At that point, the Delaware River cuts straight

The Delaware Water Gap cuts the mountain in two.

through Kittatinny Mountain, and runs along the narrow canyon-like valley floor it carved out.

If a stream no longer flows through the gap, it is known as a *wind gap*. Cumberland Gap, in the Cumberland Mountains is a good example.

In colonial days, the Cumberland Gap helped open the west to the settlers who were then living along the Atlantic coast. The people were blocked from moving westward by the Appalachians, a string of mountains too difficult to cross.

Daniel Boone, the great frontiersman, wanted to find a way over these mountains. One day, in 1769, he and three other men were roaming along the frontier near the point where Tennessee, Virginia and Kentucky come together. After passing the last settlement, they came to a low, narrow opening or gap in the mountains. The men hacked at trees, branches and underbrush. Following the natural gap through the mountain, they cleared the way to the other side.

The route they blazed through the Cumberland Mountains became known as the Wilderness Road. Settlers, traders and trappers followed the trail into the rich farmlands of Kentucky and the Ohio Valley. Today, highways and railways take advantage of gaps in crossing mountains.

Passes and gaps have other names, too. They are known as notch in New England and clove in New York. In other parts of the country, the terms, break and narrows, mean the same thing—a gap.

6

Islands

An *island* is a piece of land surrounded by water. Islands are found in lakes, rivers, coastal waters and far out in the ocean.

One day, a young child from desert country saw, for the first time, some small islands in a large lake. Perplexed, and perhaps a little worried, she asked, "Do they float?"

Islands don't float. They are formed by the forces of erosion and deposition of sediment, and by the change of sea level. Islands are:

- tops of mountains or tops of volcanoes.
- parts of the continental shelf sticking up above water.
- offshore bars and hammocks built above water level.
- formed by coral, tiny sea animals that live in large colonies. When they die, their shells stick to the ocean floor. Layer upon layer builds up and reaches above sea level.

Wizard Island, a volcano within a volcano, in Crater Lake, Oregon.

An island may be either:

- found along the shore or inland.
- alone or one of a large group.
- small as a city block or as large as a continent.
- mountainous or flat as a pancake.
- covered with trees or barren of all vegetation.
- covered with ice and snow or kept warm by the sun.
- crowded with people, shops and factories or wild and uninhabited.

Islands differ. There is only one thing an island must be. An island must be a piece of land surrounded by water.

← A few of the Thousand Islands in the St. Lawrence River between the United States and Canada.

7

Caves and Caverns

In an old story, Aladdin stood in front of a mountain, rubbed his magic lamp and called, "Open Sesame." A huge slab of rock moved, and Aladdin found himself at the opening of a cave. Curious, he went in to explore and found many treasures.

In another story, Tom Sawyer and Huckleberry Finn, quite by accident came upon a cave. They couldn't resist going in to explore it.

People who explore caves for fun are called "spelunkers." Aladdin, Tom and Huck are story-tale spelunkers. But from the beginning of time, people and animals couldn't resist going into caves. Caves have been used as homes and shelters, hideouts for outlaws, secret meeting places and tourist attractions. And caves still are used in these ways.

A *cave* is a large, natural, hollowed out space on the side of a cliff or in the earth. It may or may not have an opening large enough for a person to enter.

There are many kinds of caves:

- wind caves in the sides of mountains. Wind and grinding sand hollow out openings in soft spots on the slopes.
- ice caves in glaciers. Over and over, ice melts below the surface of the glacier and flows off until a cavity is formed.
- lava caves near volcanoes. In a lava flow, the lava hardens on the outside surface first. But inside, the hot, liquid lava keeps on flowing until it flows out beneath the hardened top, leaving a tunnel—a lava cave.
- sea caves in seaside cliffs. Waves pound on weak spots, and break away pieces of rock until an opening is made.
- limestone caves. Most caves in the United States are made of limestone, a rock that contains minerals which dissolve easily in water.

Limestone caves start when one of the gases in the air, carbon dioxide, combines with water to form a weak acid. Water as a weak acid can dissolve some minerals. The water with its carbon dioxide soaks into cracks and joints of a mass of limestone. The liquid dissolves the limestone and carries it away. The principal mineral in limestone is known as calcium carbonate. This is the mineral that dissolves and mixes with the water as it flows through the cave. With this mineral gone, the limestone is weakened. It crumbles and is carried away, too. First holes are hollowed out. Then an underground room forms. Other rooms follow. In time,

An ice cave.

passageways connect one room with the other, often at different levels.

Drops of water containing dissolved calcium carbonate continue to seep through openings in cave ceilings and walls. When the

water dries, calcium carbonate is left, forming fanciful shapes. The most familiar are the stalactites which hang like icicles from the ceiling. Dripping water from stalactites slowly builds up icicle-shaped carbonate rocks on the floor. These are called stalagmites.

Stalactites grow downward to join stalagmites to form a fantastic "stage curtain."

Sometimes, a stalactite joins a stalagmite to form a column.

A limestone cave may have either:

- a wide entrance, a narrow one, or just a slot in the earth.
- some light or may be in total darkness.
- one room or many chambers connected by tunnels.
- one level or many levels, each going deeper and deeper into the earth.
- dry rooms, damp rooms or both.
- underground streams running through them, pools captured in the rocks or none of these.
- no other caves near it or it may be part of a large, complicated cave system. (Kentucky, New Mexico, Missouri and central Florida).
- temporary guests such as bears, foxes and people or permanent residents such as beetles, spiders, snails, bats, both seeing and blind types of fish.

A cavern is a large cave. Often the two words, cave and cavern, are interchanged.

Carlsbad Caverns in New Mexico are the largest limestone caves in the world. About 37 miles of the caverns have been explored, but no one knows how far they extend. At one point, there is a drop to the depth of about 1,100 feet. One room has a ceiling 285 feet high. And some rooms are as much as 220 feet

The Giant Dome is one of the millions of fascinating formations to be seen in Carlsbad Caverns National Park in southern New Mexico. →

across. The caverns are filled with fanciful shapes sculptured by water and air.

For hundreds of years, Indians lived in the Carlsbad Caverns. Their cooking pots, paintings and drawings have been found there. Today, this underground wonderland is part of our national parks system.

8

Glaciers

A glacier is a moving mass of ice. It may move down a mountain as a valley glacier or over the land as an ice sheet.

Valley glaciers start on the tops of high mountains where there is snow all the time, even in the summer. In pockets just below the highest peaks, snow accumulates in snow fields. Each year, snow melts and freezes, but some remains. Each year, fresh snow falls.

When the pile is deep enough, the snow and ice underneath are pressed together and forced into less space. Near the bottom of the pile, the snow and ice change into a rough, grainy, icy material called firn.

Layer upon layer of firn and snow build up. Under the pres-

Yentna Glacier in Alaska. From the left, an ice fall is joining the glacier.

sure of the weight of the pile, the air between the granules in the lower layers is squeezed out. Solid ice forms.

When the pile gets too heavy, a tongue of solid, rock-like ice sticks out from the bottom of the mass. It spreads out and moves forward very slowly, carrying along the upper layers of firn and snow. The ice flows down hill like a river. On its way down, the glacier may follow a valley carved out by a river.

When a glacier follows a river valley, it fills the valley and even climbs hundreds of feet up the valley walls. It splits, gouges, scrapes and picks up rock from the floor. It breaks off sharp edged rocks from the walls. The rock is carried along in mounds, lines and ridges. All the rock moved by a glacier is called moraine.

A large glacier may be joined by smaller ones coming out of side valleys. Most valley glaciers inch their way down. But some move many feet each day. Usually, the steeper the slope the faster the movement.

As it travels, the glacier gets thinner and thinner until, at a warmer level, it melts away. At the end of its journey, the glacier leaves behind the pile of rocks it has been carrying. The rocks may be dumped in an unsorted pile, or dropped in lines and ridges going in the same direction as the glacier. These deposits are called moraines, too. (Moraine has two meanings. A moraine is the ridge or mound of rock carried by a glacier. It is also the deposit of material left on the ground by a glacier.)

An ice sheet is another type of glacier. Larger than a valley glacier, it moves over the land. While a valley glacier is long, the ice sheet is round or slightly oval.

Moraine dams a lake.

In cold lands where snow falls all the time, vast areas of land are covered by glacial ice thousands of feet thick. Near the center, an ice sheet can be 10,000 or more feet thick. The center is always thicker than the edges. The ice spreads outward from the center in all directions, north, east, south and west, very much like a mound of ice cream melting on a plate.

During the Ice Age, ice sheets moved over the continents. In North America, they moved south from Canada and covered the northern part of the United States. Ice sheets eroded the land, rounded off tops of mountains and widened valleys. When the ice retreated, moraine, about 100 feet high and several miles wide, was dropped in the form of low ridges. They stretched from Long Island through Ohio. Cape Cod in Massachusetts was also formed by moraine dropped during the Ice Age.

In some places, the ice gouged holes in rocks and scooped out deep trenches. When the ice sheets melted, these holes were filled with water and became lakes and rivers. That is how many rivers and the Great Lakes were formed. Today, ice sheets are found in the coldest parts of the earth, near the North and South Poles.

Ice caps are smaller ice sheets. In cold lands, they lie on tops of a group of mountains. From an ice cap, many valley glaciers move down mountain slopes. Today, an ice cap covers all of the continent of Antarctica.

When a glacier comes to an end near the ocean, it breaks off, drops into the water and floats away as an iceberg. This is called calving. In some places, a glacier may extend far out to sea, and float above the water as an ice shelf.

Glaciers erode the land in one place and build it up in another.

When glaciers erode the land, they:

- break off pieces from mountain tops and form jagged, saw-toothed peaks.
- round off tops of mountains.
- scoop out bowl shaped basins called cirques on the side of mountains.
- widen and flatten old valley floors.
- carve out new valleys.
- carry off and later deposit rocks, boulders and soil.
- bulldoze and grind up soil.
- grind down highlands, making them lower.
- smooth and polish stones and cut grooves into the sides of some of them.
- dig out basins that later hold lakes and ponds.

When glaciers build up the land, they:

- fashion beaches, capes, coastlines and coastal plains along the shore.
- fill valleys, sometimes several hundred feet deep.
- deposit rich, fertile soil on plains.
- deposit moraines.
- drop sediment into the sea, which in time may be used in mountain building.

Though glaciers vary, all glaciers change the face of the earth.

9

Deserts

Though deserts differ widely, one ruling applies to all. To be called a desert, a region must have less than 10 inches of rain each year. Except for the icy wastelands of the Arctic and Antarctic, deserts are hot and dry. On account of scanty rainfall, the land has little or no vegetation. Often, it is deeply eroded.

In the United States, a desert may be either:

- stony and pebbly, wind swept with little or no sand or it may be sandy with dunes moving endlessly over it.
- covered with a surface of salt or lava.
- at high altitudes or below sea level.
- a valley, mountain or plain.

- a high plateau deeply eroded into canyons and sculptured rock shapes or a low piece of land surrounded by high, steep, rugged mountains.
- barren, have some plant and animal life of its own or be reclaimed and irrigated for farmland.

There is a spaciousness about most deserts. The air is clear, and you can see for 20 to 30 miles. Mountains can be far off, but because of the clear air, they seem to appear closer.

Changes in the weather are extreme. Days are hot, nights are cold. Summers are hot, and winters often freezing. Though rainstorms are rare, they are sudden and heavy. And because the land has been starved for water, the rain disappears into the ground as quickly as it appeared. When the ground is hard, the rain water may stay on the surface.

Sometimes at a distance, a pool of water seems to cover the desert floor. It is not water but a mirage, a trick played on the eyes by the conditions of the atmosphere in the desert.

Plants and animals that need little water live in some deserts. They adapt themselves to the scanty rainfall and the long periods of heat and drought. Shrubs, like sagebrush, grow long roots which go far underground to get to water. Plants, like cacti, store water in their waxy leaves and stems. Coarse grasses grow in low bunches here and there. Animals get their moisture from plant juices or from the animals they eat. To get away from the heat, many small animals burrow underground and come out at night to feed.

PLANTS AND ANIMALS OF THE DESERT

Flowering Yucca plant

Organ Pipe Cactus

Kangaroo rat

The American deserts extend from the Rocky Mountains to the Sierra Nevada Mountains, and from central Oregon and Idaho to Mexico. They vary in altitude from highlands to lowlands.

Death Valley in California is a well known desert, 280 feet

Two views of Death Valley. (Below) The hard floor was beaten with sledge hammers to make a road. (Following pages) Rimmed by high mountains, sand dunes move restlessly over the desert floor.

below sea level. Its flat floor is covered with sand, salt and rock. Sandy dunes move over it. Nearby mountains rise steeply 11,000 feet. Summer temperatures are among the highest in the world. Though hot in summer, in winter it is pleasant. And during that season, the desert attracts many tourists. For hundreds of years, the Panamint Indians lived in Death Valley. They still do.

In the hot, dry highland, the mountains and plateaus are deeply cut by streams that follow occasional bursts of rain. Canyons and cliffs form. Later, they are eroded into mesas and buttes.

Arches, rock bridges and grotesquely shaped rocks are also carved out, and become part of the desert scenery. The torrential rains are the master sculptors of these rock forms. The water carves out weak spots in the rock, and creates fantastic shapes. Wind and sand help. The wind hurls sand against the face of the rocks and cuts out more soft spots. As time goes by, the sand smoothes and polishes the rough edges of the rock sculptures into soft curves. Some rocks look like angels, some have the shape of huge hammers, and others look like sentinels guarding the secret laboratory of a mad scientist.

Eventually, the rock shapes are broken into small bits and blown away. The highland desert becomes flat.

Wind, water and weathering carve fantastic shapes in highland deserts. →

Index

M

N

O

P

Q

R

S

Picture Credits

Corps of Engineers, U. S. Army: p. 48
Florida Department of Commerce: pp. 44-45, 54, 55
Georgia Bureau of Industry and Trade: p. 18
Georgia Department of Community Development: p. 51
William Jaber: pp. 12, 15, 21, 26, 58
Kansas State Historical Society: p. 43 (bottom)
Massachusetts Department of Commerce & Development: p. 22
Missouri Division of Tourism: pp. 98-99
New Mexico Department of Development: pp. 59, 101
New York State Department of Commerce: pp. 10 (upper left), 11 (upper left), 71, 92
Oregon State Highway: pp. 37, 78
Palisades Interstate Parkway: p. 57
Santa Fe Railway: pp. 11 (upper right), 40-41
U. S. Department of the Interior, Bureau of Land Management: p. 35 (top)
U. S. Department of the Interior, National Park Service: pp. 10 (bottom), 19, 23, 24, 28, 29, 31, 33, 38, 60, 65, 67, 75, 81, 83, 85, 87, 96-97, 104, 111, 112 (top), 113, 114-115, 117
U. S. Forest Service: pp. 50, 80
U. S. Geology Survey: pp. 35 (bottom), 69, 72-73, 90-91, 106
Vermont Agency of Development and Community Affairs: p. 63
West Virginia Department of Commerce: pp. 10 (upper right), 66